TOTAL ECLIPSE
OF THE SUN

in Cornwall and South Devon

Wednesday 11th August 1999

by

Pam Hine

YARNER PRESS

The publishers gratefully acknowledge NASA/Goddard Space Flight Center for
the maps appearing (with minor modifications) on pages 3 and 9, and for the
use of their extensive data on which much of this book is based.

ISBN 0–9533769–0–7

First Edition August 1998

Published by Yarner Press
an imprint of Richard Hine & Co. Ltd.
18 Droridge, Dartington
Totnes, Devon TQ9 6JQ

If you have difficulty obtaining further copies of this book, they may be
ordered direct from the publisher. Please send a cheque payable to
Richard Hine & Co. Ltd. for £3.50 including postage and packing.

Printed in Devon

Contents

Maps

Timetables

Thanks to my husband, Richard, to my parents, Norman and Joan Rycroft,
my stepdaughter, Abigail, and to Mike Bailey, Mary Bartlett,
Gwen Butcher, Jeff Harvey, Richard and Saj Heming,
Nigel and Viv Hinks and Carole Powell for their help and encouragement.

Thanks also to Fred Espenak of NASA/Goddard Flight Space Center.

Additional thanks to Gwen for her splendid dragon on page 19.

*"These late eclipses in the sun and moon
portend no good to us: though the wisdom of nature can
reason it thus and thus, yet nature finds itself scourged by
the sequent effects: love cools, friendship falls off,
brothers divide. In cities, mutinies; in countries, discord;
in palaces, treason; and the bond cracked 'twixt son and
father . . ."*

*"This is the excellent foppery of the world, that
when we are sick in fortune – often the surfeit of our own
behaviour – we make guilty of our disasters the sun, the
moon and the stars: as if we were villains by necessity,
fools by heavenly compulsion . . ."*

William Shakespeare
King Lear, Act I Sc ii

A PREDICTION

It is predicted that in the middle of the morning of Wednesday, 11th August 1999, darkness will descend in parts of Cornwall and Devon as the Sun gradually becomes a smaller and smaller crescent of light and eventually disappears altogether. Birds will stop singing and animals and plants will behave as if night is coming.

Fortunately, it is also predicted that, after a couple of minutes of darkness, the Sun will gradually reappear and normal life will resume.

In ancient times, the event would have been heralded as a sign of the gods' displeasure or as the precursor of catastrophe. However, through the scientific study of the solar system, we know today that this extraordinary phenomenon is actually a total eclipse of the Sun.

WHAT IS AN ECLIPSE?

An eclipse of the Sun happens when the Moon moves between the Earth and the Sun, so that the three bodies are in line. Viewed from Earth, the Moon masks the Sun. If the Moon covers the Sun completely, it is a total eclipse. If the Sun is only partly obscured, it is a partial eclipse.

By a remarkable coincidence, the Sun and the Moon appear to be almost the same size when viewed from the Earth. The orbits of the Moon around the Earth and the Earth around the Sun are elliptical rather than circular. Because of this, their apparent sizes change. Sometimes the Moon appears to be smaller than the Sun and sometimes larger. It is only when the apparent size of the Moon is larger than that of the Sun that a total eclipse is possible.

We are all aware of the motion of the Sun across the sky. Day after day it rises in the East and sets in the West. Most of us are much less aware of the motion of the Moon. In fact it follows a very similar path to that of the Sun but is slightly slower, taking just under 25 hours to make a complete circuit of the Earth. Although we may not notice the Moon very often during daylight, it is present in the daytime sky for, on average, half of the time. Once every 29 days or so, at the time of the New Moon, the Sun overtakes the Moon, usually passing above or below it. When the Sun passes behind it, we have an eclipse. We cannot normally see the New Moon because we are looking at its dark side.

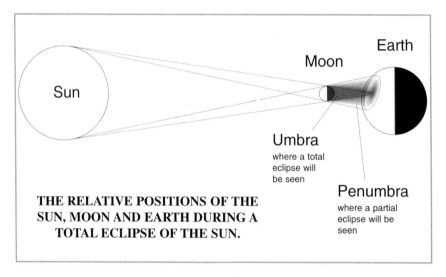

THE RELATIVE POSITIONS OF THE SUN, MOON AND EARTH DURING A TOTAL ECLIPSE OF THE SUN.

The central area of dark shadow, from which a total eclipse is seen, is called the 'umbra'. The much larger area of shadow surrounding the umbra is called the 'penumbra', from which a partial eclipse is seen. Within the penumbra, the depth of shadow reduces gradually from the edge of the umbra to the outside edge, reflecting the extent to which the Sun is covered by the Moon.

The 'path of totality' is the path which the umbra follows as it travels across the face of the Earth.

VIEW OF THE WORLD SHOWING THE PATH OF TOTALITY AND AREAS OF PARTIAL ECLIPSE ON 11TH AUGUST 1999

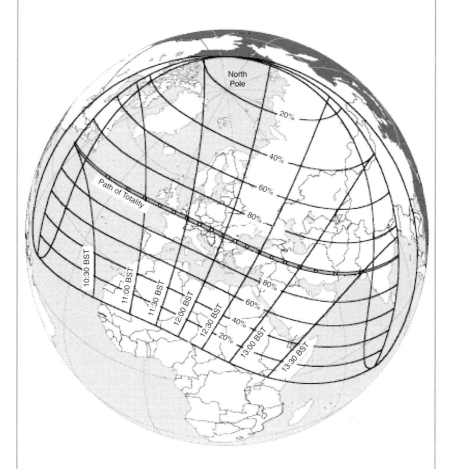

The path of totality stretches from the western Atlantic across Europe to India. The lines running parallel to this path show the extent of the penumbra and the percentage of the Sun's diameter which is obscured when the eclipse is at its maximum. The time lines at half hour intervals give the time when the eclipse is at its maximum.

The orbital plane of the Moon around the Earth lies at an angle of about 5° to the orbital plane of the Earth around the Sun. For the three bodies to line up sufficiently for an eclipse, the Moon must be close to the line of intersection of these planes. Because of this, eclipses are relatively rare.

A solar eclipse (eclipse of the Sun) happens when the Moon is towards the Sun and a lunar eclipse (eclipse of the Moon) happens at the time of Full Moon when the Sun and Moon are on opposite sides of the Earth. A total lunar eclipse is seen much more often than a total solar eclipse – firstly because the Earth casts a much larger shadow than the Moon and secondly because a lunar eclipse can be seen by everyone on the dark side of the Earth.

A total eclipse of the Sun occurs on average about once every 18 months. However, any one point on the Earth will be in the path of totality on average only once every 360 years. On the UK mainland the last total solar eclipse was in 1927, in the north of England. The next will not happen until the year 2090.

IN THE PATH OF TOTALITY

On the morning of 11th August 1999, weather permitting, people in Cornwall and South Devon will see a spectacular phenomenon, probably the most awe-inspiring light show in the world.

Shortly before 10 o'clock the partial phase will begin. Taking care to look through protective viewers (see page 16 for more information), we will see a small bite taken out of the top right hand side of the Sun. Very slowly, over the next hour or so the bite will become bigger and bigger and the light level will reduce.

It will get darker more rapidly as the crescent Sun diminishes to a fine line. Refraction in the Earth's atmosphere may then cause 'shadow bands' which are faint ripples of light, best seen against a light background. It may feel as if a bad storm is approaching but with no sign of dark clouds, it can be a bit disconcerting. An eerie

STAGES OF THE ECLIPSE SEEN FROM THE PATH OF TOTALITY

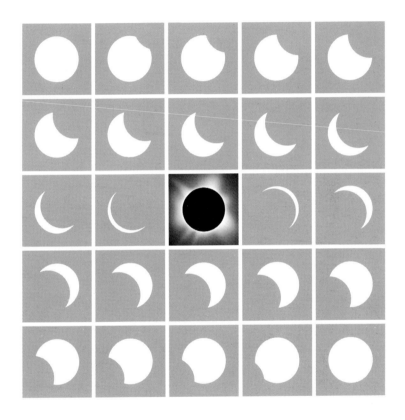

This is how the Sun will appear over a two and a half hour period on the morning of 11th August 1999. The Sun travels across the sky faster than the Moon and passes behind the Moon as it overtakes.

silence will descend as animals and birds behave as if it were nightfall. Flowers may begin to close their petals.

As the umbra approaches, we should see a darkening of the sky on the western horizon.

The Sun's crescent will reduce to a string of bright beads of light as the last rays shine between the mountains of the Moon. These are known as Baily's Beads.

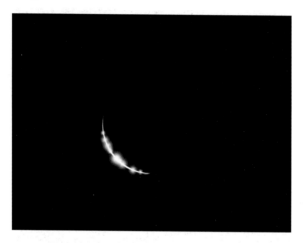

One by one the beads will disappear. When only one remains, we will see the stunning effect known as the 'diamond ring'.

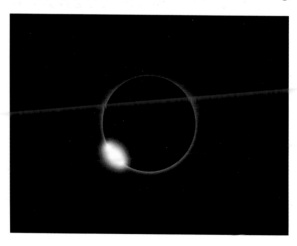

When the last bead has gone, it will be safe to look towards the Sun with the naked eye. We will be faced with an awesome and extraordinary sight: a large black disc where the Sun should be and around it the Sun's magnificent corona, like a giant cosmic flower, extending several Sun-diameters out into space. This has been called the 'Eye of God'.

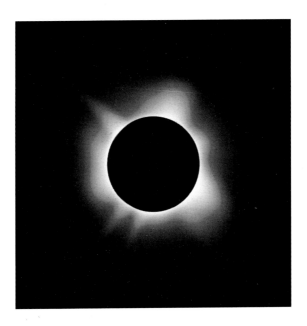

At the edge of the Moon will be seen the Sun's chromosphere (literally sphere of colour) appearing briefly as a thin crescent of red light. The occasional prominence may protrude from the Sun's surface – binoculars or a telescope will give a better view.

Stars and planets will be visible in the middle of the day – Mercury will appear to the west of the Sun and Venus to the east. There may be meteor showers in the western sky.

When Baily's Beads begin to reappear, the magical phase of totality is over and it is time to replace your eye protection.

Your relationship with the Sun will never be the same again!

Only people within the path of totality will see a total eclipse of the Sun. The path starts at dawn in the Atlantic Ocean off the eastern coast of the USA. This is when the umbra first touches the Earth. It then travels across the Atlantic Ocean, reaching the Isles of Scilly off the south-west tip of Cornwall at about 11.10 am (see map opposite).

The path of totality over the south-west of the UK will be just over 100 kilometres (65 miles) wide. Travelling at about 3200 kph (2000 mph), the umbra will take just 5 minutes to cross Cornwall and South Devon.

The northern limit of totality is a line drawn from Port Isaac on the north coast of Cornwall to Teignmouth on the east coast of Devon. The main centres of population in the path include Penzance, Truro, Plymouth and Torquay (see pages 10 and 11 for timetables).

Leaving the east coast of Devon, the umbra goes rushing off across the English Channel, including Alderney in its track. It then travels across northern France, central Europe and the Middle East. Finally it moves over India and the path finishes in the Bay of Bengal at sunset as the shadow of the Moon hurtles off into space.

ON THE EDGE – THE 'ZONE OF GRAZING'

The 'zone of grazing' is a strip about 2 km (1 mile) wide at the edge of the path of totality. Observers in this zone should see an excellent display of Baily's Beads, as the edge of the Moon grazes past the edge of the Sun. However, they will miss out on totality.

It has been calculated that the southern part of Teignmouth, Bishopsteignton and Kingsteignton lie within the zone of grazing. No doubt many people from these places and others who live in or close to the zone will get on their bikes to go south into the path of totality.

PATH OF TOTALITY OVER SW ENGLAND AND NW FRANCE

The oval shapes represent the outline of the umbra at five minute intervals

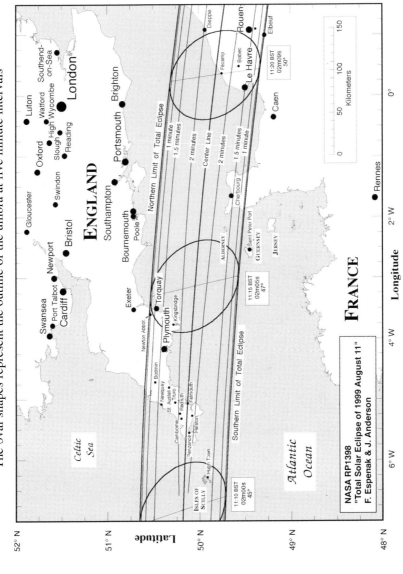

TIMETABLE FOR CORNWALL

Location	Partial Eclipse Begins (BST) hour:min	Totality Begins (BST) hour:min:sec	Duration of Totality min:sec	Partial Eclipse Ends (BST) hour:min
Bodmin	09:58	11:12:20	1:25	12:33
Camborne	09:57	11:11:00	2:00	12:32
Falmouth	09:57	11:11:20	2:05	12:32
Fowey	09:58	11:12:05	1:50	12:33
Hayle	09:57	11.10.50	2:05	12:32
Helston	09:57	11:11:00	2:05	12:32
Land's End	09:56	11:10:20	2:00	12:31
Liskeard	09:58	11:12:35	1:25	12:33
Looe	09:58	11:12:20	1:50	12:33
Mevagissey	09:57	11:11:45	2:00	12:33
Newquay	09:57	11:11:35	1:40	12:32
Padstow	09:58	11:12:10	1:05	12:33
Penzance	09:57	11:10:35	2:05	12:32
Port Isaac	09:58	11:12:45	0.25	12:33
Redruth	09:57	11:11:10	2:00	12:32
St. Agnes	09:57	11:11:15	1:55	12:32
St. Austell	09:58	11.12.00	1:50	12:33
St. Ives	09:57	11:10:45	2:05	12:32
Truro	09:57	11:11:25	2:00	12:32
Wadebridge	09:58	11:12:15	1:10	12:33

TIMETABLE FOR SOUTH DEVON

Location	Partial Eclipse Begins (BST) hour:min	Totality Begins (BST) hour:min:sec	Duration of Totality min:sec	Partial Eclipse Ends (BST) hour:min
Ashburton	09:59	11:13:55	0:55	12:35
Brixham	09:59	11:13:50	1:30	12:35
Buckfastleigh	09:59	11:13:40	1:10	12:34
Dartmeet	09:59	11:13:55	0:35	12:34
Dartmouth	09:59	11:13:40	1:40	12:35
Ivybridge	09:58	11:13:15	1:35	12:34
Kingsbridge	09:59	11:13:15	1:55	12:35
Newton Abbot	09:59	11:14:15	0:35	12:35
Plymouth	09:58	11:12:50	1:40	12:34
Prawle Point	09:58	11:13:10	2:00	12:35
Salcombe	09:58	11:13:10	2:00	12:35
South Brent	09:59	11:13:30	1:25	12:35
Tavistock	09:58	11:13:30	0:40	12:34
Teignmouth	09:59	See page 8	See page 8	12:35
Torquay	09:59	11:14:05	1:10	12:35
Totnes	09:59	11:13:40	1:25	12:35

TIMETABLE FOR THE ISLANDS

Scilly Isles	09:56	11:09:35	1.46	12:30
Alderney	10:00	11:15:15	1.47	12:37

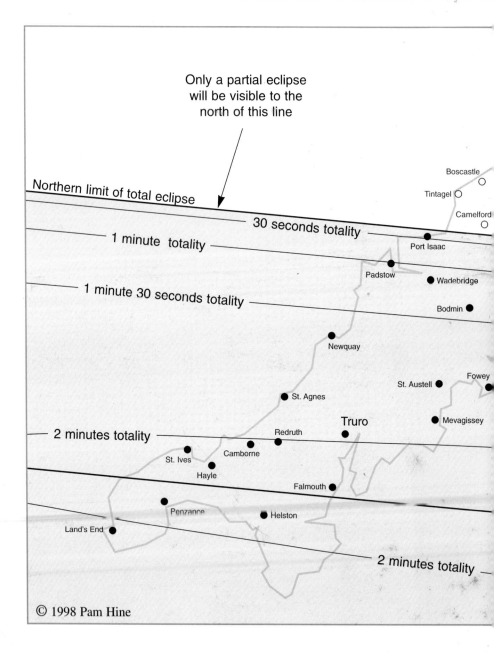

Only a partial eclipse
will be visible to the
north of this line

Northern limit of total eclipse

30 seconds totality

1 minute totality

1 minute 30 seconds totality

Boscastle ○

Tintagel ○

Camelford ○

Port Isaac ●

Padstow ●

Wadebridge ●

Bodmin ●

Newquay ●

Fowey ●

St. Austell ●

St. Agnes ●

Truro ●

Mevagissey ●

2 minutes totality

Redruth ●

Camborne ●

St. Ives ●

Hayle ●

Falmouth ●

Penzance ●

Helston ●

Land's End ●

2 minutes totality

© 1998 Pam Hine

12

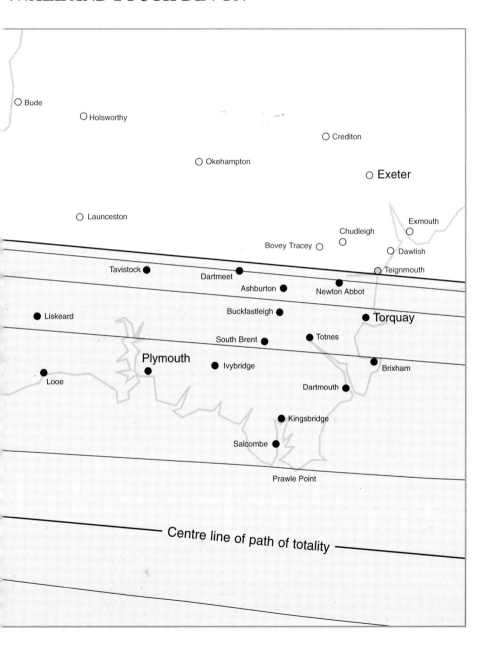

IN THE AREA OF PARTIAL ECLIPSE

Everyone in the UK will see a partial eclipse of the Sun which will last for about two and a half hours. The maximum proportion of the Sun obscured, and consequently the reduction in daylight, depends on how far north you will be. People in London will see up to 96.6% of the Sun obscured, while those in Edinburgh will see up to 81.9% obscured (see table below for more details of these and other places). Wherever you are in the UK, it will be a very special event. People will remember it for the rest of their lives.

TIMETABLE FOR UK PARTIAL ECLIPSE			
Location	Partial Eclipse Begins (BST) hour:min	Partial Eclipse Ends (BST) hour:min	Maximum Proportion of Sun Obscured
Aberdeen	10:08	12:34	77.6%
Belfast	10:01	12:30	86.9%
Birmingham	10:02	12:36	93.5%
Bristol	10:00	12:36	97.3%
Cardiff	10:00	12:35	97.2%
Edinburgh	10:05	12:33	81.9%
Exeter	09:59	12:34	99.7%
London	10:03	12:39	96.6%
Manchester	10:03	12:35	90.1%
Newcastle	10:06	12:36	84.9%
Nottingham	10:04	12:36	91.7%
Southampton	10:01	12:38	98.8%

For the rest of the world, the area of partial eclipse extends from the North Pole to central Africa and covers parts of North America, all of Europe and much of Asia (see map on page 3).

MAXIMUM ECLIPSE VALUES FOR THE UK

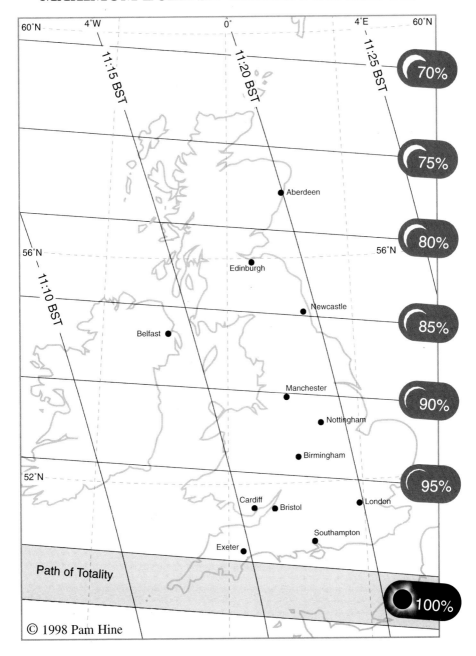

© 1998 Pam Hine

EYE SAFETY

- **Never look directly at the Sun with the naked eye.**

- **Never look at the Sun through a camera, telescope or binoculars without correct filters – seek specialist advice.**

- **The only time that it is safe to look towards the Sun with the naked eye or through binoculars, etc. is during the period of totality when the Sun is completely covered by the Moon.**

- **Sunglasses do not give protection against the damaging effects of ultraviolet and infrared light.**

Safe ways to view the Sun include:

- Special viewers with black polymer or aluminised mylar lenses. Viewers should bear the CE mark as your assurance of safety. The aluminised mylar type suffer from having a highly reflective surface on both sides, which means that you are looking through a reflection of your own eyes. The black polymer type give a much better view of the Sun. *(These may be obtained from: TRIUS, 266 Dartmouth Rd, Paignton, Devon TQ4 6LH Tel. 01803 844101.)*

- Viewing through No.14 welders glass.

Permanent damage or blindness can be caused by looking directly at the Sun, or by using unsuitable viewing materials.

ALTERNATIVE WAYS OF OBSERVING THE SUN

The following methods are good for keeping track of the progress of the eclipse. You might like to try them in advance of the big day. The projected images of the Sun will normally be circular. During a partial eclipse, the images will of course be crescent-shaped.

* Take a hand mirror and a piece of card big enough to cover it. Cut a hole in the card about 15 mm in diameter and attach it to the front of the mirror with Blu-Tack or sticky tape, forming a small round mirror. Hold the mirror in full sunshine and direct the reflected beam onto a shaded wall about 10 metres away. Fix the mirror in position so that you can have a closer look at the projected image, which should be about 90 mm in diameter. A smaller hole in the card gives a fainter but less fuzzy image which can best be seen if the image is projected into a garage, darkened room or large cardboard box. This method is really good for children and groups.

* Take two pieces of white cardboard. Fix the first piece so that it is approximately perpendicular to the Sun. Make a pinhole in the other piece and hold it about a metre away, so that the Sun's rays

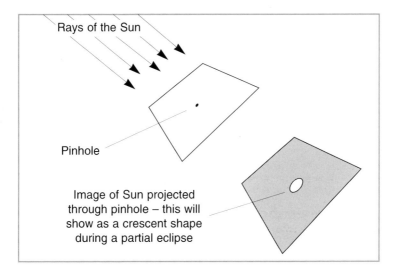

Rays of the Sun

Pinhole

Image of Sun projected through pinhole – this will show as a crescent shape during a partial eclipse

shine through the hole. Move this card until you see an image of the Sun projected onto the first piece of card.

* When you are next under a tree which produces dappled shade, look for elliptical pools of light amongst the shadows. Gaps between the leaves act as pinholes, projecting images of the Sun onto the ground. You won't always see them but keep looking – it is a delight when you do see the effect.

MYTHS AND LEGENDS

To ancient peoples the Sun was the source of all light and warmth and hence the source of all life. It was regarded as a deity and worshipped, often holding a leading place in their mythologies. Any apparent threat to this giver-of-all-life was taken very seriously. To many peoples, an eclipse was a sign of the displeasure of the gods. To others, it foretold some great disaster such as war, pestilence, famine or the death of kings.

In ancient Mexico, they thought that an eclipse signified a quarrel between the Sun and the Moon, whereas in Tahiti they thought that the Sun and Moon were making love.

The Inuit believed that during an eclipse the Sun and the Moon temporarily left their place in the heavens to check that things were all right on Earth.

A story from Africa goes something like this: In olden days the Moon was very pale and dull and was jealous of the Sun with his glittering fine feathers of light and fire. One day, the Moon took advantage of a moment when the Sun was looking at the other side of the Earth to steal some feathers of fire for herself. But the Sun found out and in his anger splashed the Moon with mud which remains stuck to her forever. Ever since that time, the Moon has been seeking revenge. Sometimes, she surprises the Sun and splashes him with mud. Then the Sun cannot shine for a while and it makes the whole world sad.

In many cultures, it was thought that during an eclipse the Sun was being devoured by a dragon, or by some other mythological being. People would do whatever they could to scare the monster away and restore daylight. They shouted. They beat drums and kettles. They hit their dogs to make them howl. They pinched their babies to make them cry. They fired burning arrows into the air at the invisible beast.

And, of course, these measures do work. The Sun has always returned – so far!

TRANSPORT MATTERS

The last total eclipse of the Sun in the UK was in 1927. It saw the biggest-ever movement of people in peacetime – a reported three million travelled to the north-west of England to witness it.

People wishing to see the total eclipse in August 1999 who do not live in the path of totality will have to travel. Many from the eastern side of the UK will go to western Europe – or even further for the better weather prospects.

Estimates vary for the number of eclipse-watchers who will want to come to Cornwall and South Devon, but it could be millions. Travel will be very difficult for some days before and after the eclipse. Excursions by plane or train are likely to be booked well in advance – do not expect to pop down to Cornwall or Devon for the day unless you happen to have your own private helicopter.

In a normal week in August, the peak holiday season, the area has about half a million visitors with most of the holiday accommodation full. If you are intending to visit in August 1999, you would be wise to plan carefully how and when you are going to travel and where you are going to stay.

If you choose to drive to the West Country, you are advised to carry plenty of supplies for the trip, especially water, as there may be long delays. Plan to travel well in advance of the day – anyone still stuck in a traffic jam to the north of a line joining Teignmouth to Port Isaac on the morning of the eclipse will be very disappointed. You might think of bringing bicycles with you – it may be the best way to move around.

Road and rail travel present special problems because the area where totality will be seen is at the end of a long peninsula and can only be approached from one direction. Lateral thinking and careful planning on the part of the authorities will be needed to move the expected numbers of people into the area, look after them while they are here and move them out again. Talks began in 1996 and the emergency services have already given a lot of thought to the particular challenges which they will face.

If you live in the West Country, the influx of visitors on top of those normally expected will have a significant effect on everyday life. Roads will be even more congested than in a typical August. Shops and other businesses may experience difficulties with deliveries and with staff getting to and from work.

Seasoned eclipse-watchers say that the duration of totality makes little difference to the experience – the important thing is to be within the path of totality. Travelling any distance on the day for the sake of a few more seconds of totality will almost certainly not be worth the hassle.

Many people will take to the sea in boats, large and small. There is a huge area of the English Channel and the Atlantic Ocean from which a total eclipse will be seen (see map on page 9) but most people will probably congregate at the major resorts. Even the water is going to be crowded.

And all for a two-minute light show.

PHOTOGRAPHY

Advice from experienced eclipse-watchers is that if this is your first eclipse, just watch and enjoy it. Photographing an eclipse is an art and unless you really know what you are doing, results can be very disappointing. You might have no worthwhile pictures after spending the whole time looking through the viewfinder – when you could have been absorbing the experience. There will be loads of excellent photographs published by the experts after the event.

If you are still keen to take pictures or video, look at the references on pages 27 and 28. Whatever you do, remember to disable any flash guns. Flash gives no advantage and would greatly upset your fellow eclipse-watchers.

Never look at the Sun through a camera without correct filters.

WEATHER PROSPECTS

The weather in the south-west of England can be unpredictable in August and local conditions may vary. From past weather records, the probability of seeing the eclipse is estimated to be about 45%.

As if the British weather was not problem enough, an eclipse of the Sun can generate its own weather. As the shadow of the Moon moves across the surface of the Earth the heat from the Sun is blocked out causing a significant drop in temperature. In certain conditions, water vapour may then condense to form clouds.

Whatever the weather, anyone within the path of totality will have the uncanny experience of a period of darkness in the middle of the day. The really spectacular sights will only be seen if the clouds stay away. Keep your fingers crossed for clear skies.

ODDS, ENDS AND FOOD FOR THOUGHT

* The expected huge increase in the number of people in the South-West will put unprecedented demands on resources and services – water, electricity, gas, sewage disposal, shops, restaurants, food, etc.

* Organisers of conferences, courses, business meetings and other events which are to be held in the South-West during August 1999 will need to consider carefully the implications of the eclipse and its possible effects on life in the area.

* People who live in Cornwall and Devon should be aware that if they leave, perhaps to go on holiday, and wish to return in early August 1999, they will face many of the same travel problems as visitors. Likewise if they're planning to travel soon after the eclipse.

* Special campsite 'villages' are planned to accommodate up to ten thousand people each.

*   The longest duration of totality for any solar eclipse is just over seven minutes when viewed from the ground. In 1973 totality lasted much longer for passengers on a special Concorde flight which stayed in the Moon's umbra for a record 74 minutes.

You may have already missed your chance to book a seat on Concorde as it tries to race the eclipse in 1999. It will rendezvous with the umbra over the Atlantic Ocean, but even travelling at 2700 kph (1700 mph) it will soon be overtaken.

*   The QE2 and other cruise liners are scheduled to come into Falmouth. With all the boats sounding their sirens and hooting their horns, Falmouth may not be the quietest place to enjoy the eclipse. It should be a lot of fun, though.

*   If it is cloudy, you will be able to see the eclipse on television as live coverage will be transmitted from planes flying above the clouds.

*   Street lights will switch on automatically as light levels fall. They are individually activated by light sensors at the top of each lamp-post.

*   In some countries the police close all the roads for a total eclipse of the sun – it may happen here!

*   If you get bitten by the bug and want to join the ranks of the eclipse-chasers, the next total solar eclipse will be in Africa on 21st June 2001. The path of totality starts in the southern Atlantic Ocean, crosses Africa from Angola to Mozambique, and after passing over Madagascar ends in the Indian Ocean.

*   The next total lunar eclipse will be in the early hours of 21st January 2000. It will be visible from the whole of the UK.

Jargon Buster

Baily's Beads

Just before totality, the last rays of the Sun shining between the mountains of the Moon appear as a bright string of beads. These are known as Baily's Beads after the English astronomer, Francis Baily, who described the effect in 1836. They can be seen again as the Sun reappears after totality.

Chromosphere

(literally sphere of colour). The lower atmosphere of the Sun. It appears as a thin red crescent for a few seconds at the beginning and end of totality.

Corona

The outermost regions of the Sun's atmosphere which have a temperature of around two million degrees C. During totality, the visible area appears as a halo around the Sun.

Partial Eclipse

An eclipse as seen from the penumbra (partial shadow). An observer sees the Sun reduced to a crescent-shape as it is partially obscured by the Moon.

Path of Totality

The path followed by the umbra as it travels across the face of the Earth.

Penumbra

The area of partial shadow of the Moon where only part of the light from the Sun is blocked out. An observer in the penumbra sees a partial eclipse of the Sun.

Prominences

These are flame-coloured projections of hot ionised gas rising from the surface of the Sun. We should see them during the total eclipse.

Total Eclipse

An eclipse during which the umbra touches the Earth. An observer within the umbra sees the Sun totally obscured by the Moon.

Totality

The time when the Sun is completely covered by the Moon.

Umbra

The part of the Moon's shadow where all of the light from the Sun is blocked out. An observer in the umbra sees a total eclipse of the Sun.

A FEW DIMENSIONS

To give you an idea of the relative sizes of the Sun, the Earth and the Moon, imagine a football pitch.

In one goal-mouth is the Sun, a giant beach-ball 1 metre (3 ft) in diameter. In the other goal-mouth is a pea. A peppercorn is circling the pea at a distance of about 300 mm (1 ft). The pea is the Earth and the peppercorn is the Moon.

In reality, the sizes are a bit bigger:

- The diameter of the Sun is 1,392,000 km (865,000 miles).

- The diameter of the Earth is 12,740 km (7,920 miles).

- The diameter of the Moon is 3,480 km (2,160 miles).

- The Earth is on average 150,000,000 km (93,000,000 miles) from the Sun.

- The Moon is on average 384,400 km (238,860 miles) from the Earth.

The ratios of the diameters of Sun:Earth:Moon are about 400:4:1.

FURTHER READING AND RESEARCH

Books

The Royal Greenwich Observatory Guide to the 1999 Total Eclipse of the Sun by Steve Bell.

Eclipse! The What, Where, When, Why and How Guide to Watching Solar and Lunar Eclipses by Philip S. Harrington.

Eclipse by Bryan Brewer.

UK Solar Eclipses from Year 1: An Anthology of 3,000 Years of Solar Eclipses by Sheridan Williams.

Chasing the Shadow: Observer's Guide to Eclipses by Joel Harris and Richard Talcott.

Guide to Solar and Lunar Eclipses by Alex Vincent.

Historical Eclipses and Earth's Rotation by F. R. Stephenson.

Total Eclipses of the Sun by J. B. Zirker.

Wonders of the Sky: Observing Rainbows, Comets, Eclipses, the Stars and Other Phenomena by Fred Schaaf.

The Cambridge Eclipse Photography Guide by J. Pasachoff and M. Covington

And for Children:

Nature's Blackouts by Billy Aronson

Eclipse Information on the Internet

A good starting point is:

http://www.hermit.org/Eclipse1999/

which gives lots of information and links to many other relevant sites including all the incredibly extensive information supplied by NASA at:

http://umbra.nascom.nasa.gov/eclipse/990811/rp.html

Another excellent NASA site shows a series of photographs as the umbra crosses the Earth during the total eclipse of February 1998. The photographs were taken from the Geosynchronous Operational Environmental Satellite (GOES). They can be found at:

ftp://climate.gsfc.nasa.gov/pub/chesters/goes/980226.eclipse/gif/980226.eclipse.tiles.gif

The Solar Eclipse 1999 UK Co-ordinating Group was formed to :

- Perform useful scientific experiments on the solar corona.
- Use the eclipse to increase public understanding and appreciation of Science & Technology.
- Help the public to get the most out of the eclipse.
- Act as a focal point for co-ordination of activities leading up to the eclipse.

Their site is:

http://ast.star.rl.ac.uk/eclipse99/index.html

Eclipse Photography on the Internet

Lots of expert advice can be found at:

http://planets.gsfc.nasa.gov/eclipse/sephoto.html